Poultry Housing

by Professor James Dryden

with an introduction by Jackson Chambers

Self Reliance Books

Get more historic titles on animal and stock breeding, gardening and old fashioned skills by visiting us at:

Introduction

I am pleased to present yet another title on Poultry.

The work is in the Public Domain and is re-printed here in accordance with Federal Laws.

As with all reprinted books of this age that are intended to perfectly reproduce the original edition, considerable pains and effort had to be undertaken to correct fading and sometimes outright damage to existing proofs of this title. At times, this task is quite monumental, requiring an almost total "rebuilding" of some pages from digital proofs of multiple copies. Despite this, imperfections still sometimes exist in the final proof and may detract from the visual appearance of the text.

I hope you enjoy reading this book as much as I enjoyed making it available to readers again.

Jackson Chambers

Poultry Housing

By

JAMES DRYDEN

The problem of good housing of fowls is closely related to the maintenance of health and vigor in the flock. In the past twenty years ideas have changed considerably in regard to poultry-house construction. Formerly, as now, the main consideration was comfort and health of the fowls, but comfort is now secured by building houses on a different plan from those of twenty years ago or more. The change has been one of the two or three important factors responsible for the development of a profitable poultry husbandry.

In the old plan of housing, the comfort of the fowls was obtained or sought by building warm houses, so-called, which included double-boarded walls and sufficient glass windows to let in a little light and sunshine. They were usually closed up tight in the winter time, and the result was damp, cold houses. The sun during the day, striking through the windows, caused the temperature to run up. The same glass windows were responsible for a severe drop in temperature during the night. Not the high or low temperature, but the wide range of temperature between day and night was responsible for the failure of this kind of house. With a high temperature during the day, the air took up much moisture, and at night the temperature dropped so much that the moisture in the air was condensed on the walls and in cold sections froze, making the house far from comfortable. This kind of house offered a favorable environment for various poultry diseases, and the diseases did not miss the opportunity.

This type of house, however, has now been scrapped, and the comfort and health of the fowls is secured by building, not the so-called warm houses, but rather cold houses that maintain a better equilibrium between night and day temperatures. Glass windows are largely discarded, and the double wall and dead air space idea is now pretty well a relic of the past. The modern idea is an open-front house, and it is this type of house that has brought success to poultry keepers. The satisfactory result has come chiefly through attention to ventilation.

LOCATION

(1) **Soils.** Chickens will thrive on a great variety of soils, but certain kinds are more adapted than others to successful poultry keeping. Especially in choosing a location for commercial poultry farming, heavy clay soils should be avoided, if possible, as they are hard to keep clean and sanitary. A rather light porous soil is preferable; this is drier in wet weather and not as hard in dry weather. A wet soil is colder than a dry one.

(2) **Drainage.** If the ground selected has not good natural drainage, provision should be made either by under drainage or by open

ditches for carrying off the surplus water. The water should not be allowed to stand in the yards. Muddy feet mean muddy eggs. Dampness means catarrh, roup, rheumatism, tuberculosis, etc.

(3) **Air Drainage.** Air drainage is sometimes as important as soil drainage. Cold moist air seeks the lower levels. It is better to locate the house and yards on higher levels where there is some air movement to carry off the cold, damp air or prevent its becoming stagnant. Fowls should not, however, be exposed to high winds. You will notice that on windy days the fowls mope around in sheltered corners or in houses. This is not favorable to high egg production. Sufficient air drainage without interfering with the comfort or activity of the hen is the ideal condition. The houses may be built on the leeward side of an orchard or in the shelter of buildings. A windbreak of trees may be set out to provide shelter.

(4) **Sunshine.** If possible the houses and yards should be built where they will get full benefit of the sunshine. Face them south unless the prevailing winds are from that direction. Sunshine is a germ destroyer and a better egg producer than red pepper or other condimental foods.

(5) **Other Points.** Other points that should be considered in locating the houses are: (a) The convenience of the attendant. Nearness to the feed and water supply will save labor. (b) Building the houses away from the other buildings. This will make it easier to keep the premises free from insect pests, rats, etc. (c) A location near a city market or a good shipping point, other things being equal, should be given the preference.

REQUIREMENTS

No one style of poultry house will suit all conditions; besides, fowls may be kept successfully in different kinds of houses. There are certain principles, however, that should govern in the building of any kind of poultry house.

(1) **Ventilation.** Fowls require considerably more fresh air than other farm animals. It has been estimated that a hen in proportion to her weight requires double the weight of oxygen that a man or a horse requires. The amount of air breathed per 1000 pounds live weight of hens is given by King as 8,272 cubic feet in 24 hours, the requirement of a man being 2,833 and a cow 2,804 cubic feet. Ventilation, moreover, keeps the house dry. A close, warmly built house, with glass windows, is always damp, because of the extremes of temperature between day and night. Dampness is overcome by ventilation. Ventilation can best be furnished by leaving one end or one side open or partly open. In cold sections or exposed or windy locations this opening may be covered with a curtain of muslin; this will admit air, yet add to the comfort of the fowls. In most sections, however, the curtain will not be needed. The other three sides of the house should be draft-proof. The idea is to give plenty of ventilation without drafts.

(2) **Protection from Cold and Storms.** A good poultry house should afford protection from the winds and the storms and severe

weather. A little shelter from the winds and the storms will increase egg yields. A cold wave or a sudden change to colder weather means an immediate demand for increased fuel to keep up the heat of the body. In this case the fuel is the food that the hen eats, and the food that has been going into the making of eggs will be drawn upon for fuel purposes. Any shelter, therefore, that protects fowls from storms or sudden changes in temperature is an incentive to egg production. Fowls maintain rugged health roosting in trees, but the sudden and frequent changes in the weather to which they are subject in the trees interfere with egg production.

(3) **Space Required per Fowl.** The house, while affording protection from storms, should also furnish scratching room, for activity is the life of the hen. Where there is little or no snow or where the chickens can be out of doors almost every day in the year, about two square feet of floor space for each fowl will be sufficient. This will apply to flocks of twenty or more hens. For smaller flocks a more liberal allowance of space should be made. Where the climate is such that the fowls will seek shelter part of the year rather than go out of doors in the yards and fields, more floor space should be provided, say three to four square feet for each fowl. The house should be built high enough for a man to work in without bumping his head. This height will allow sufficient air space for the fowls.

(4) **Floors.** Floors are not always necessary in poultry houses. A floor will be an advantage where the ground is inclined to be damp, but where it is well drained and porous there need be no floor. A wooden floor gives protection for rats underneath, and for this reason a cement floor is preferable. On the other hand, cement floors do not permit moving the house, where this is desired. Where an earth floor is used it should be higher than the ground outside of the house, to prevent water running in. It is a good plan to fill the floor with six or eight inches of clean, coarse sand, and once a year or oftener to scrape off part of this and replace it with clean sand. This will keep the floor comparatively clean and lessen the danger from diseases in the flock.

THE PORTABLE HOUSE

The portable house is used where the colony system prevails. Much of the trouble from diseases comes from keeping the chickens on the same ground year after year. By keeping them on clean ground, which is possible with portable houses, they are under natural and hygienic

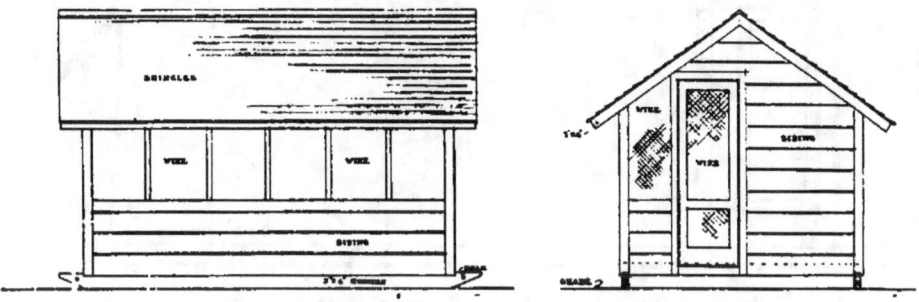

Fig. 1. Side and end elevation of O. A. C. portable colony house.

conditions. This system, moreover, fits in with a system of crop rotation on the farm. About fifty chickens to the acre will keep the land in high fertility. Besides, the chickens will find a considerable portion of their food in the waste grain and weed seeds, grasshoppers, and other insects. They often rid the farm of grasshoppers and other insects, thus saving valuable crops. Another important advantage of the colony system, is the fact that the fowls are more active when they have the liberty of fresh fields than when confined in yards. Finally, with the colony system

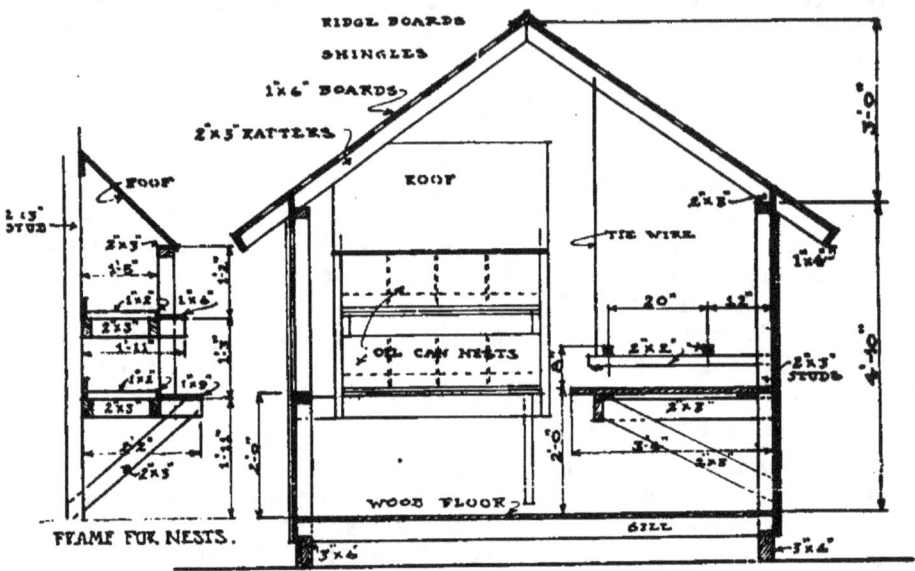

Fig. 2. Plan of frame for nests and end section of O. A. C. portable colony house.

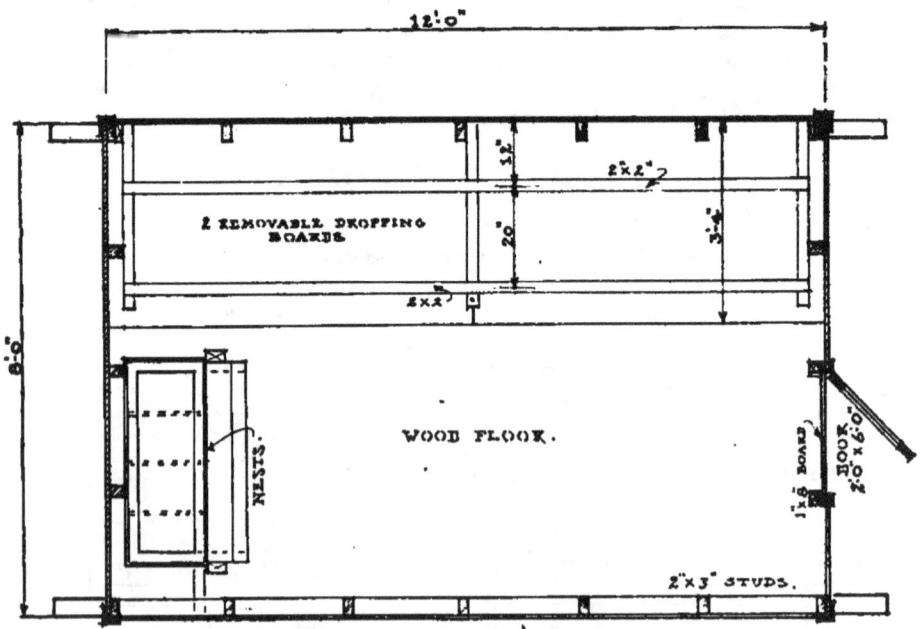

Fig. 3. Floor plan of O. A. C. portable colony house.

there is no expense for fencing. If fowls are kept on an extensive scale, where much cheap land is available, the colony system is undoubtedly the best, but with high priced land any system of fifty hens to the acre is out of the question.

A good size for a portable house is 8 by 12 feet, which will accommodate 40 fowls. A team of horses will pull a house of this size, which is built on runners. It may be used as a stationary house.

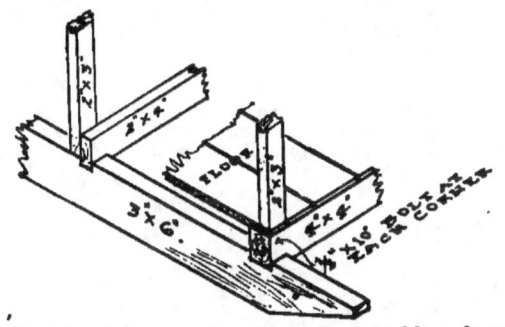

Fig. 4. Detail of runners for O. A. C. portable colony house.

DETAILS OF CONSTRUCTION

Bill of Materials for Portable Colony House 8'x12'

Lumber.

2	8"x6"	14 feet long	runners.
2	4"x4"	8 feet long	sills.
5	2"x4"	8 feet long	sills.
14	2"x3"	5 feet long	studs.
4	2"x3"	7 feet long	studs.
2	2"x3"	8 feet long	studs.
3	2"x3"	12 feet long	plates.
14	2"x3"	6 feet long	rafters.
8	2"x3"	12 feet long	nest frames, etc.
2	2"x2"	12 feet long	roosts.
3	2"x2"	3 feet long	roost supports.
175	board feet	8" lap for flooring and dropping boards.	
125	board feet	6" roosting boards and slats for dropping boards.	
260	board feet	8" channel rustic siding, No. 2.	

1250 shingles.

4	1"x4"	corner boards, each 6 feet long.	
4	1"x3"	corner boards, each 6 feet long.	
5	1"x3"	door and door opening. Each 12 feet long.	
2	1"x4"	14 feet long cornice finish.	
4	1"x4"	6 feet long cornice finish.	
1	1"x8"	14 feet long ridge board.	
1	1"x4"	14 feet long ridge board.	
5	1"x3"	door and door opening. Each 12 feet long.	
1	1"x2"	14 feet long stops for oil can nests.	

Hardware.

6 lbs. 8 D case nails.
10 lbs. 8 D common.
3 lbs. 16 D common.
4 lbs. shingle nails.
1 pair of strap hinges.
6 feet of heavy wire.
18 feet of 3' mesh wire for door and front.
8 10" 1x10" x 15" oil cans for nests. 2/3 of one end cut out.
4 ½"x10" anchor bolts.

.A 100-HEN HOUSE

A plan of a good house of 100-hen capacity is given herewith. A cement floor, on the whole, is the most satisfactory in a permanent house, but it may not be necessary in some locations.

BILL OF MATERIAL

Concrete work. (1:2.5:5 mixture. 1:2 wearing surface.)

    10 barrels of Portland cement.
    2 cubic yards of sand.
    5 cubic yards of gravel.

Lumber.

    7 4"x4" 6 feet long.
   14 2"x4" 10 feet long.
   12 2"x4" 12 feet long.
    4 2"x2" 12 feet long.
    2 2"x2" 14 feet long.
  125 board feet of 8" ship lap for dropping boards.
  400 board feet of 8" channel rustic siding.
 3250 shingles.
   50 1"x6" 14 feet long for roof sheathing.
   32 1"x4" 14 feet long for miscellaneous.
    4 1"x5" 14 feet long for eccentric corner and ridge boards.
    2 1"x2" 10 feet long for stops for oil-can nests.

Hardware.

   18 lbs. 8 D case nails.
   18 lbs. 8 D common.
    6 lbs. 16 D common.
   10 lbs. shingle nails.
    One pair of 4" strap hinges.
    Three heavy wires six feet long each.
   30 feet of 4'-1" mesh wire.
    Six ½"x10" anchor bolts for bottom plate.
   16 10"x15" oil cans for nests; 2/3 of one end cut out.
    Paint.

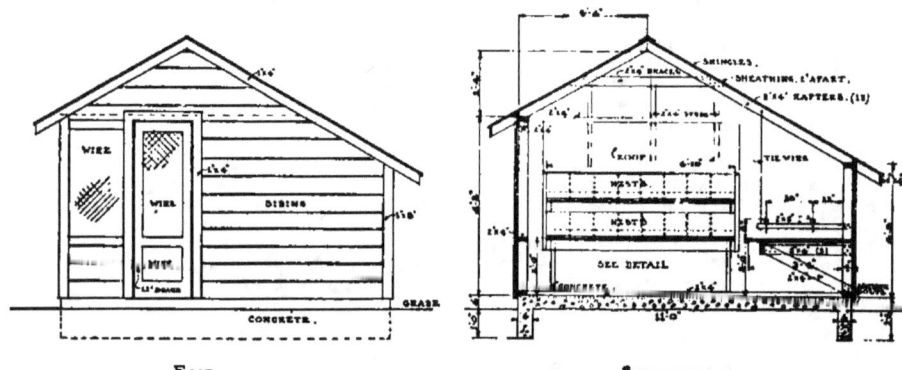

Fig. 5. End view and section of 100-hen house.

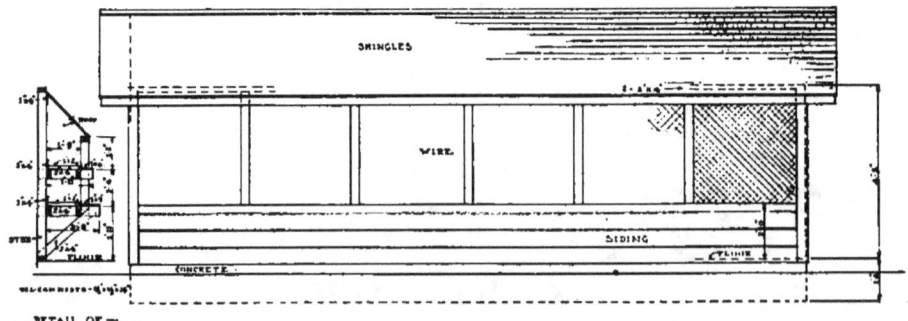

Fig. 6. Front elevation and detail of frame for nests, 100-hen house.

INTERIOR ARRANGEMENT

Whether the house be a portable farm poultry house, a breeding house, or a large commercial laying house, practically the same arrangement of the interior may be had. All the floor space should be used by the fowls for scratching. If possible all nest boxes, perches, hoppers, etc., should be movable. This will make it easier to clean out the house and to keep it free from lice and mites.

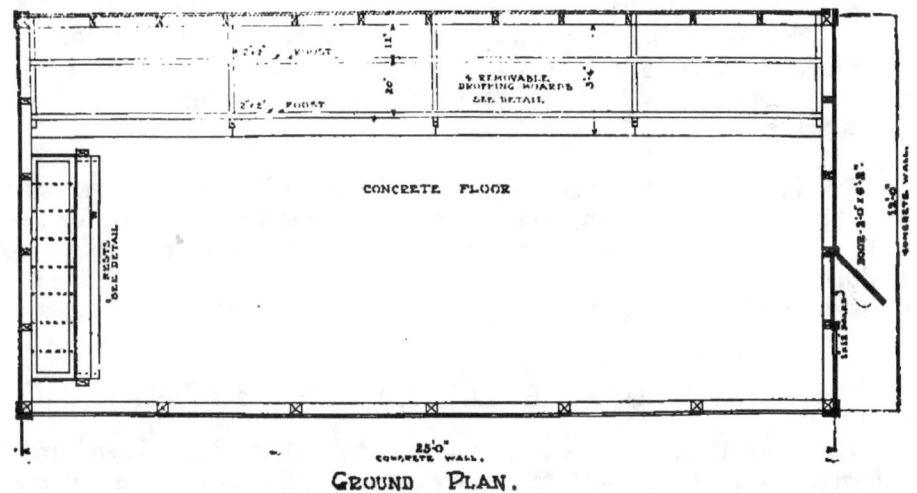

GROUND PLAN.

Fig. 7. Ground plan of stationary house.

The Nests. Nests for laying hens should be somewhat secluded, for fowls are less likely to acquire the egg-eating habit when the nests are in a darkened place. They should be from 10 by 12 to 12 by 14 inches in size and 8 to 10 inches high, the larger breeds requiring the larger size. A cheap and serviceable nest may be made out of a five-gallon oil-can by cutting the end out, leaving about three inches at the bottom to keep the nest material in the nest. Such a can may be easily cleaned either by washing or spraying. The illustration shows top of can taken off; this makes the nest more roomy. Several of these nests may be set on a platform about 2 feet from the floor, turning the entrance of the nest towards the wall and leaving a space or walk along the back. The nest platform should be about two feet from the ground, being nailed to a cleat on the side of the house and braced from top of sill. Over the nests, to keep the chickens from standing on them and to help to darken them, should be fitted a sloping top. This top should be built high enough so that the attendant can see into the nests. One nest to five or six hens should generally be provided. The plan of

Fig. 8. Good serviceable nests may be made out of 5-gal. oil-cans. Such a nest is cheap and easily cleaned. An oil-can may also be used for watering the flock.

putting the nests under the platform, shown in the 500-hen house plan, will be satisfactory for the 100-hen house as well as the portable house. These nests are made in sections of five ten-inch nests and slide under the platform. The hen enters at the rear. It is recommended that the nests be made with wire bottoms mainly because there will be fewer eggs broken. As the hen rises when she lays, the egg is likely to break when it drops on the board bottom, since straw may not always cover the bottom of the nest. The front of the nest is hinged and it is readily opened in gathering the eggs.

Broody Coop. A coop for broody hens for the commercial house is placed over the platform on each side of partition in the center of the house. This coop should be about 2 by 3 feet running back to the wall, and made of slats 1½ inches apart, with a wire bottom of 1-inch poultry netting. This coop should not interfere with the raising or lowering of the perches.

Perches. It is important that they be placed far enough apart to prevent the fowls becoming overheated and to assure fresh air for them to breathe. The fresh air house must furnish fresh air at night as well as during the day. The perches are made of 2-by-3-inch stuff. The hens roost on the 2-inch surface, which should have the sharp edge planed off.

THE COMMERCIAL LAYING HOUSE

Commercial poultry farms, which in this State have been largely a development of the past ten years, are usually located on suburban tracts of ten to twenty acres, though where large numbers of fowls are kept, this acreage must be extended. The high price of land near towns and cities emphasizes the need of economy in the amount used. In some cases, no provision for an outside yard is made, the fowls being confined in the house all of the time. The same plan is sometimes followed where there is sufficient land for outside yards, but the soil is of such a character that during the winter season the yards are wet and muddy. With such soil and restricted yards it is a question whether yards are any advantage during the rainy season, or whether the fowls are not better kept in the house all of the time.

A number of poultrymen are now keeping their laying flocks confined in houses throughout the year. Others keep them confined only during the wet season. Experiments made at this Station a few years ago indicated that a satisfactory egg yield can be secured where the fowls are confined in the house throughout the year.

These results are supported by the records of several small pens of the station flock. In 1914-15 a pen of twenty-four Oregons were kept shut in a small house 8 by 12 feet throughout the year, and they made a record of 229.16 eggs a hen. This is ten eggs a hen more than were laid by a larger pen that had the liberty of a roomy yard every day in the year.

In 1915-16 a pen of twenty-five Barred Rock pullets were confined in a house throughout the year, and they laid an average of 182.6 eggs a hen, practically the same production as our experimental flocks of the same breeding. The production was as follows by month per hen:

	1915-16	1914-15
October	....	1.66
November	1.10	6.75
December	10.10	17.45
January	13.65	18.95
February	16.85	20.54
March	22.60	23.50
April	22.80	22.95
May	20.50	21.95
June	17.45	18.54
July	17.10	19.58
August	15.10	18.79
September	12.40	16.20
October	7.90	15.20
November	4.25	6.62
December	.70	.20
Average	182.6	229.16

The annual average is for the twelve months beginning with first egg laid by each hen.

Another pen of twenty-five Oregons confined throughout the year 1915-16 in a house 10 by 14 on the writer's back-yard lot, made an average of 216 eggs a hen, though the feeding was not particularly favorable. In each case they had plenty of fresh air and were fed in such a way as to induce maximum exercise.

These three flocks were of the Station's high-producing strain. Where the fowls have plenty of exercise and right feeding there appears to be little difference in the yield, whether confined or on range. In other words, wide range out of doors is not necessarily essential to high production.

This has no relation, however, to the other problem of fertility and hatching quality of the egg. As to what effect close confinement has on the breeding quality of the stock has not yet been determined definitely, but the best information available points to the necessity of giving the breeding stock plenty of good range. The cumulative effect of continuous confinement of breeding stock, generation after generation, is likely to result in lowered vigor. Very few breeders confine breeding stock.

YARDS

If the poultry houses are located near a neighbor's field or gardens it will be necessary to yard the fowls. For other reasons, such as the keeping of more than one variety or strain of fowls, separate fenced enclosures must be maintained.

Keep the Yards Clean. When chickens are confined in yards throughout the year, care must be taken to keep the yards clean, otherwise in time there will be serious losses from diseases and general lowering of vitality. The yards sooner or later become contaminated with disease-producing germs, and losses through sickness and decreased vitality will render it unprofitable to keep fowls. Dr. Salmon says: "Accumulations of excrement harbor parasites, vitiate the atmosphere, and breed contagion." It may be possible to keep yards sufficiently clean by disinfection and other means to prevent troubles of this kind, though this is doubtful. At any rate, the expense of disinfection and cleaning would render it impracticable.

Size of Yards. The larger the yards the more exercise the fowls will take. The fowls will do better in a large yard than in a small one, even though the same number of square feet be allowed each fowl, because there is more room for exercise. Each fowl has the use of the whole yard whether it be large or small. The cost of fencing will largely govern the size of the yard. Fencing is expensive, and if the yards are very large the cost may exceed that of the houses. For a pen of 100 fowls a yard 100 by 100 feet is a good size.

Double Yards. In order to give the fowls clean ground and guard against "ground poisoning," the best plan is to have double yards; that is, two yards for each pen of fowls. If one long continuous house be used and it is divided into small pens, it will be better to have the yards on each side of the house, rather than two yards on one side, thus obtaining width enough in the yards for cultivation. Where the yards are too small for horse cultivation, spading will have to be resorted to.

Separate Houses. Under some conditions and for special breeding pens, a more convenient arrangement than one long house is to use small separate houses, placed in a row 20 feet apart. By this arrangement the yards may all be on one side of the house, and one can walk or drive a team on the other side from one end to the other without opening gates. Another advantage is that there is less danger of contagious diseases spreading from one flock to another than in the continuous house; every flock is practically isolated from the other.

Cultivation of Yards. The cultivation and cropping of the yards will keep them in good condition. The crops will use up the manure and lessen the danger from spreading of disease. The cultivation also keeps the surface of the soil loose; unless cultivated, the soil, if of clayey nature, will become hard and baked from continuous use. The expense of building the extra fence for the double yards will be offset by the value of the crops that may be grown on the vacant yards. It will pay to plow the yards two or three times during the year. Cultivation has a double purpose; first, it cleans the yards; second, it offers the fowls more exercise.

Crops to Grow. Different kinds of crops or vegetables may be grown on the vacant yards. Green food may be grown for the fowls, or vegetables may be grown for the family. The droppings of the fowls will keep the soil in a very productive condition. If it is not desired to use the yards for garden purposes, such crops as vetch, alfalfa, clover, kale, rye, etc., may be grown. Where it grows well, clover may be planted early in the spring and the chickens turned onto it in the fall. Vetch sown in the fall will furnish a great quantity of excellent green food in the spring and summer. Where it thrives, probably no other forage plant will furnish more green food to the acre than thousand-headed kale. If planted early in the spring, it will furnish a great quantity of green food in the fall and following winter. Rye sown in the fall will make considerable green food in the following spring and summer.

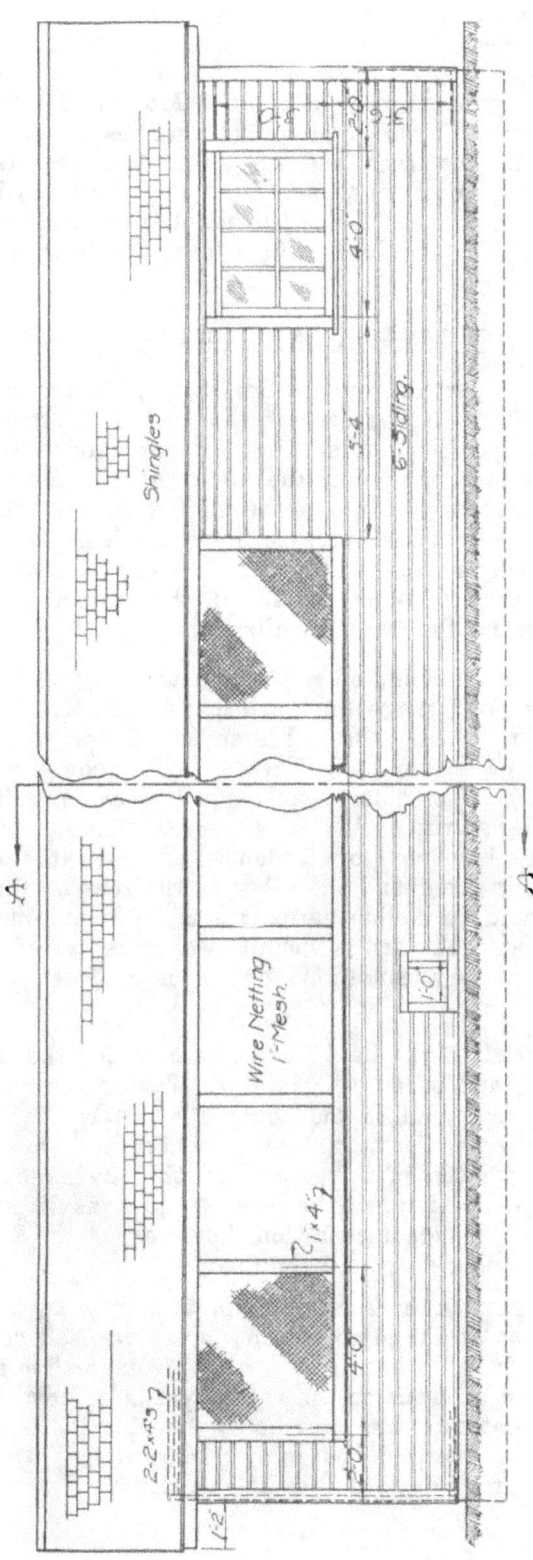

Fig. 9. Front elevation, 500-hen house.

Shade. For fowls in summer shade is very necessary. This may be secured from fruit trees or other trees. Two or three fruit trees, such trees as will do best in the particular soil and location, in each yard, will furnish some revenue as well as shade. Most varieties of fruit do well in poultry yards. Plums, apples, and cherries do especially well. The droppings fertilize the trees, and the poultry aid materially in keeping in check certain of the fruit pests. Sour apples should be fed sparingly to fowls. Where it is not desired to plant trees, sunflowers or corn may be planted early in the season in a part of the yard fenced off temporarily.

COMMERCIAL UNITS

In commercial poultry keeping a desirable plan for housing is to build in 500-hen units. As the size of the flock is increased the production is decreased per hen, other things being equal. On the other hand, the larger the unit, the lower the labor costs in feeding and general care. Under certain conditions the saving in cost in the larger units may more than offset the decrease in production. If there is good management there may be little difference in the final returns from the 500-hen units and the 1000-hen units. If the proper yard space is available we recommend the smaller units.

Range. As to the amount of range and the character of the soil, in general the more good range the fowls have the better will be their health and production. Where the soil is sufficiently dry and porous the fowls for best results should have the use of a roomy yard. For a flock of 500 hens a good plan is to set aside two acres, using one acre at a time and growing a crop on the second acre. It is not very material whether the land is all in front of the house, divided into two yards, or whether one yard is in front and the other at the rear of the house. In any event there should be double yards. A more ideal condition would be three yards so that each yard is vacant two years out of three. This will insure better sanitary conditions and greater freedom from contagious diseases.

Close Confinement. If the land lacks drainage and the soil is heavy, which means muddy, wet yards all winter in Western Oregon, it is just as well to keep the hens shut in the house all winter. The same practice should be followed in Eastern Oregon where the snow covers the ground for several months in the year. If the fowls get the proper feed and care the production will be just as good as if they had the liberty of outdoors. The ideal condition, however, if the soil is of the right character, is outdoor range all the year.

For permanent results there is no doubt that it is advisable to give the maximum amount of range consistent with overhead costs. If the land is not of too high value it will pay to give four or five acres to 500 hens. By the growing of crops on the vacant yards, the overhead charge need not be much greater for the four acres than for the two. The conditions would then be favorable not only for good egg production but also for keeping the stock in good breeding condition.

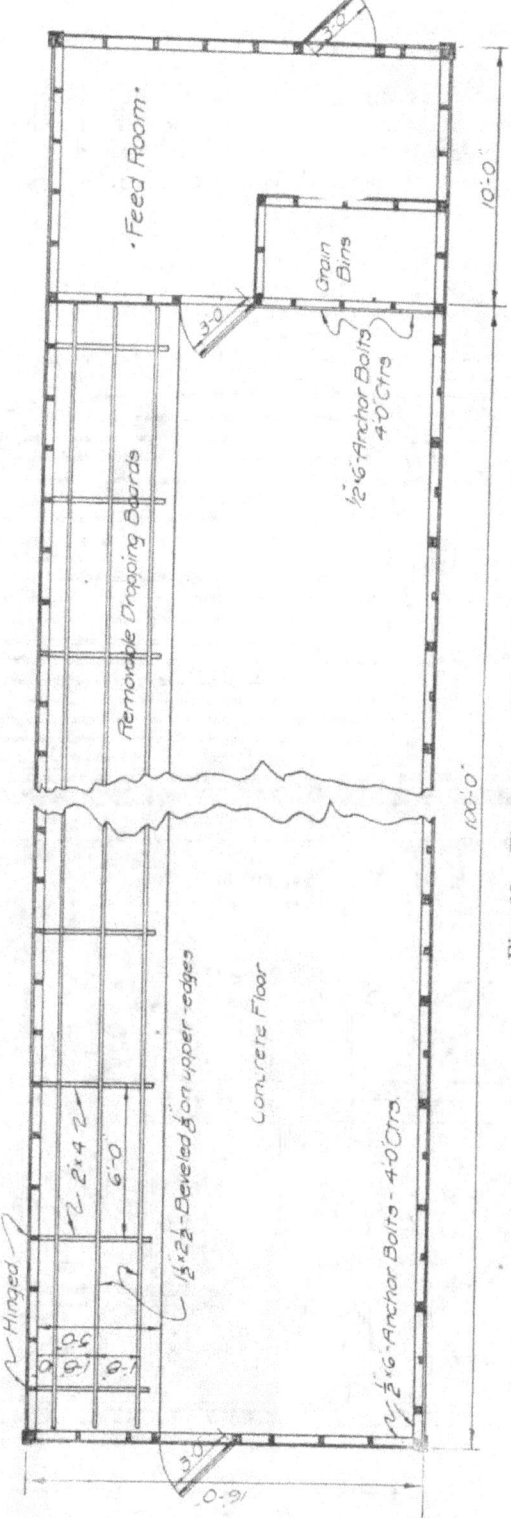

Fig. 10. Floor plan, 500-hen house.

FIVE-HUNDRED-HEN HOUSE

The style of house illustrated herewith will be suitable for Western Oregon and in general for sections of the State where the temperature does not get much below zero. There is no reason why in Western and Southern Oregon the open-front house can not be used throughout the year.

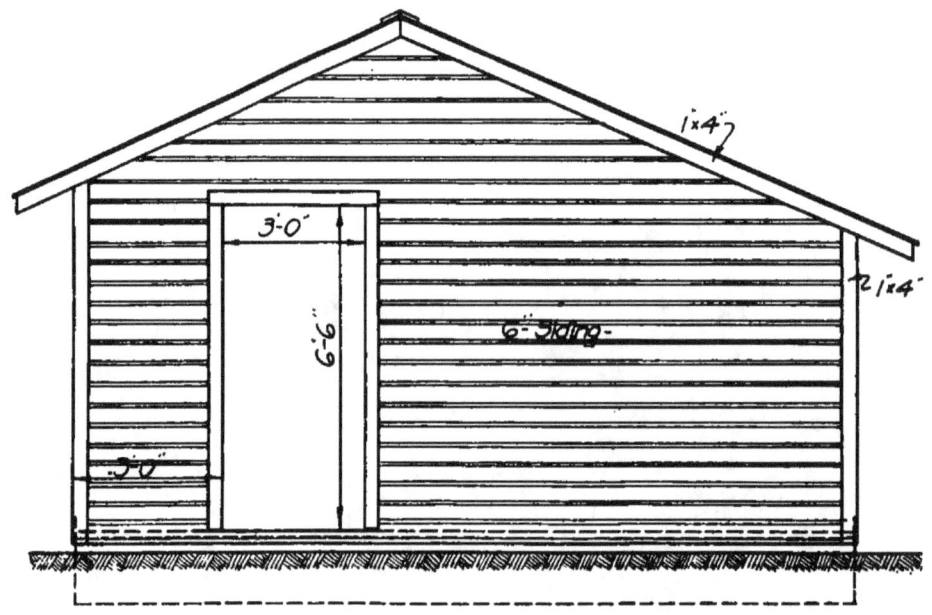

Fig. 11. End view, 500-hen house.

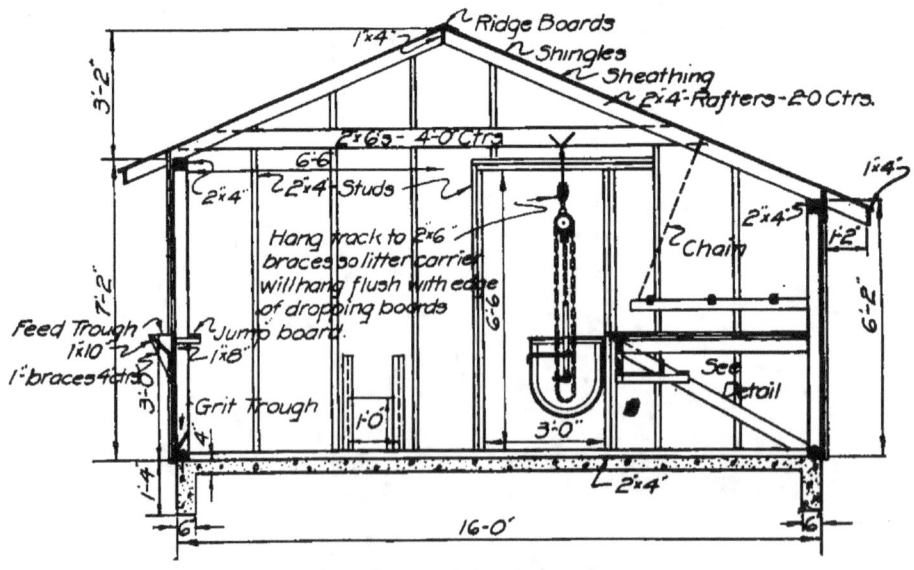

Fig. 12. Section AA, 500-hen house.

The fowls will be more comfortable on the floor if they are protected from draft in the colder months by closing the opening at the bottom of the wall three feet, but not more. If this is extended too high, there will not be sufficient light in the house. As to the size of the opening, the drawing shows a front wall of 7 feet 2 inches, which is boarded up 3 feet from the bottom, leaving 4 feet of opening. This will be ample.

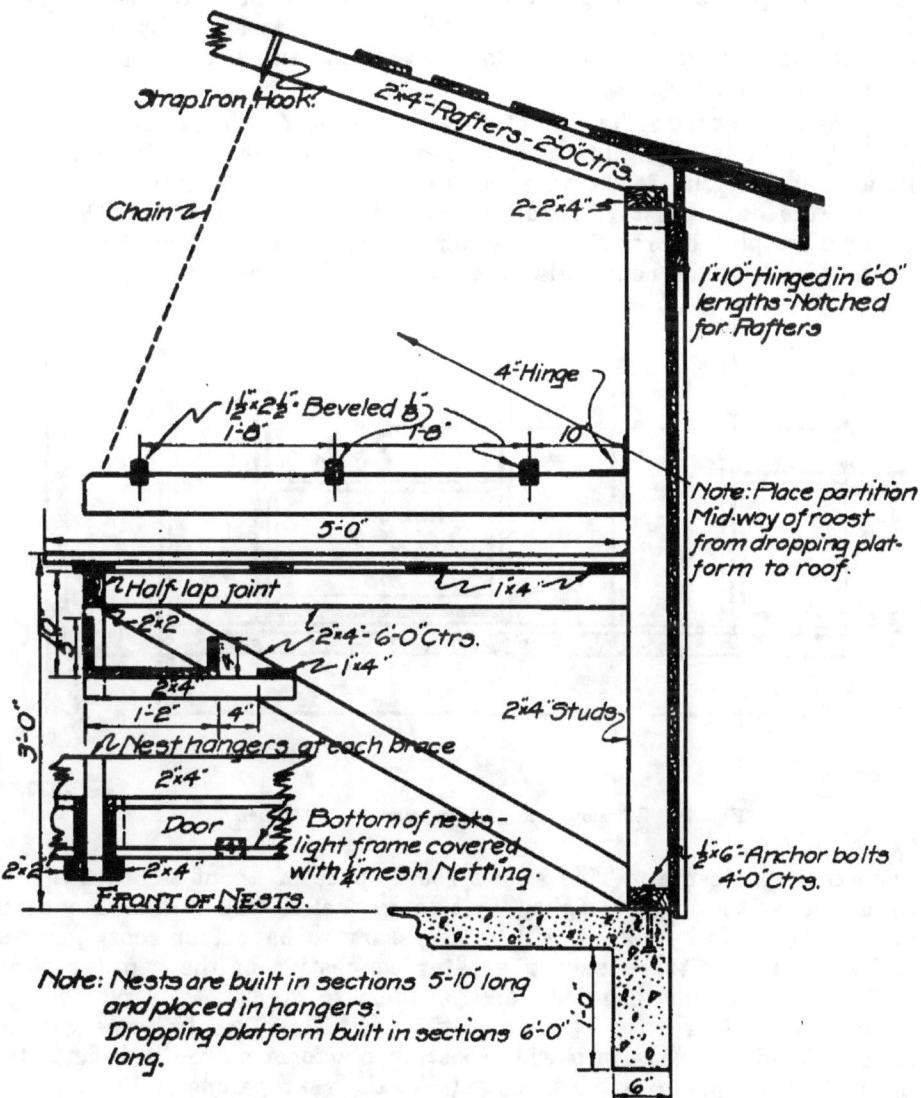

Fig. 13. Section of rear wall and roost, 500-hen house.

In the cold sections of the State, it will be better to modify the plan by placing three windows in the front, one at each end and one at the center for light and covering the rest of the opening with muslin during the cold weather. The windows should be on hinges so that they may be opened on warm days and in summer they may be removed altogether.

In Western Oregon, where the prevailing winds come from the south or southwest, the house should face the east or southeast. In other sections where the direction of these winds is different, the house may be faced toward the south.

As to the dimensions of the house, 16 feet wide and 100 feet long is a good size for 500 hens. There is no particular objection to making it 18 feet or 20 feet wide and cutting off on the length, but with a house deep in proportion to length the supply of fresh air is not so well distributed, and there is less fresh air for the fowls at night. They are crowded more together, and to that extent the full benefit of the open-front is not obtained.

The difference in the cost of construction of the house of 16-, 18- or 20-foot width is negligible and need not be considered seriously. It is important that the fowls have plenty of fresh air at night as well as during the day and this condition cannot be secured where the roost perches are placed very close together. This house calls for three perches the length of the house, placed eighteen inches apart or twenty inches

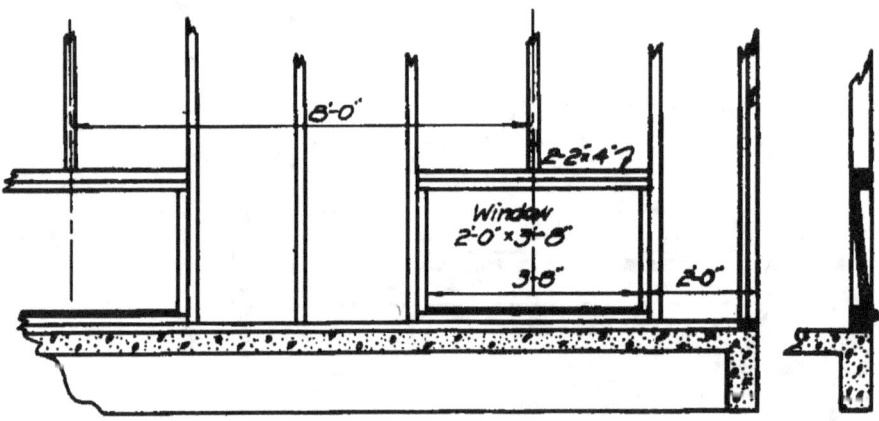

SECTION

Fig. 14. Window framing, in rear wall, 500-hen house.

from center to center. This allows roost space of about 7 inches a hen. In a twenty-foot house eighty feet long, which would give the same floor space as 16 by 100, it will be necessary to have four roost perches instead of three; this means a greater congestion of the hens at night and consequently more impure air for them to breathe. By placing the roosts far enough apart, this objection will be offset to some extent. Again, a wide house with a wide roosting platform makes it difficult for the poultry keeper to catch the fowls on the rear perches.

Rear Windows. Under certain conditions, windows in the rear of the house under the roost platform are an advantage, since they light up the floor better. It is asserted that such windows are necessary so as to prevent the fowls from scratching the litter under the platform, the theory being that a hen when scratching turns her head to the light. Details for these windows are shown in the plans. Their use is left optional. Where the house faces the east and is well lighted from the front, there are some disadvantages in the windows. With windows in the west side, the sun shining through them in the summer days will make the house uncomfortably warm. This objection may be obviated to some extent by leaving the windows partly open and by whitewashing the windows. There is also the extra cost of the windows. The argument that without the windows the litter will pile up under the roost platform at the rear is of no great importance. With a well-lighted floor and proper care in throwing the grain in the litter, there will be no trouble from this source. Where the house faces south, there will be no objection to the rear windows except the cost, and under certain conditions where the light is not good the windows will be an advantage.

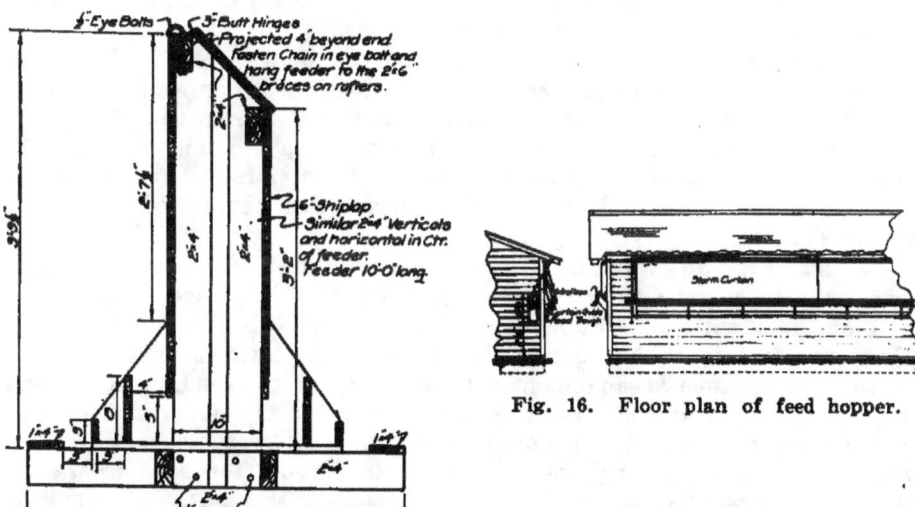

Fig. 15. Section of feed hopper.

Fig. 16. Floor plan of feed hopper.

Rear Ventilation. Provision is made in the drawing for ventilation over the roost platform above the plate by means of hinged 1-by-10 inch boards in 6-foot lengths, notched for the rafters. These boards should be closed during cold weather and open only during warm weather.

The Roof. In the drawing a shingle roof is shown; this is the most permanent type of roof. A good quality of roofing paper, also satisfactory, costs less than shingles. With paper, a flatter roof is just as satisfactory as the steeper roof necessary for shingles.

Feed Hopper. A dry mash hopper, ten feet long and ten inches wide, is detailed. This is suspended lengthwise of the house, from the 2-by-6-inch ties for rafters. The sides of this hopper are vertical and the hens eat from both sides. The whole hopper is used for dry mash and there are no partitions in it. A trough for grit, charcoal, shell, etc., is easily provided between the 2-by-4 studding on the three-foot front wall and just the width of the studding. This should be high enough from the floor to prevent litter being scratched into it. If the bottom is laid on cleats and the front hinged, this trough will make it more easily cleaned out.

Moist Mash Trough. Where moist mash is fed, a feeding trough may be attached to the front wall at the top. This trough may also be used for green stuff. (A good way to feed kale, however, is to hang up the whole plant in the house, on a wire suspended from the rafter or rafter braces.) The hens eat out of the trough through a graduated mesh wire used to cover the open front with the wide mesh at the bottom. This trough should be no higher than the three-foot front, the outer board being hinged at the bottom to facilitate cleaning. A four-inch jump board on which the hens stand while eating extends the length of the house. This board should be removable so that when pullets are first put into the house, the board can be taken down at night until they have become accustomed to go to the perches.

Litter Carrier. A litter carrier is shown on the plan. For large houses, this will be an advantage in removing the droppings. The carrier hangs flush with the ends of the dropping board, so that the droppings are scraped into the carrier. The track may extend out through the end of the house, the manure being dropped into a cart or into a manure shed. When not in use, the carrier can be pushed back into the feed room.

Watering. In Western Oregon, where the temperature is mild, water can be piped into the house. A good plan is to use an ordinary pail or pails with an automatic shut-off. These should be placed up about two feet from the floor in the center of a platform made of slats or wire. The same arrangement can be made for the feeding of buttermilk. In cold sections, this system cannot be used on account of the water pipes freezing in the winter time. In such sections, it will be a good plan to have a small heater in the feed room for heating water.

The Roosts. The roosts are made in a frame, hinged to the studding at the rear and suspended by chains from the rafters at the front. The platforms are in sections six feet long, and are removable.

The Nests. The plan shows detail of nest under the platform. These nests are built in sections five feet ten inches long, and placed in hangers. The bottom of the nest is made of galvanized wire cloth for fruit driers, ¼-inch mesh, in preference to boards, to prevent breakage of eggs. The hen enters the nest from the rear, and the front is closed by a board, hinged at the bottom, which is let down when the eggs are gathered. There are no partitions for separate nests. A number of hens lay together in one section. This makes a nest of simple construction, easily kept clean and free from mites.

Fig. 17. Controlling apparatus of storm curtain.

Roller Curtains. Under certain conditions, an adjustable curtain is desirable. For this purpose, a roller curtain such as that designed by Shoup of the Western Washington Station is very convenient. This curtain may be raised or lowered in a few seconds, by turning a crank at one end. A detail of the Shoup curtain is given herewith. In the mild climate of Western Oregon, the writer does not recommend the use of the curtain. With the house faced away from the prevailing winds and storms, there will seldom be any need for the curtain. In the unusual severe weather, such as that of December, 1919, a curtain would have been a decided advantage for two or three nights, but such weather may not come again in a generation. Many poultrymen avoided serious loss during the zero weather by covering up the opening with burlap or muslin. The danger from the curtain is that it may be closed when it should be open, and open when it should be closed. A stationary curtain and glass windows may be the best arrangement for the cold sections.

BILL OF MATERIAL FOR 500-HEN HOUSE

Description	Size	No. Req'd.	Description	Size	No. Req'd.
Rafters	2″x4″x12′	56	Feeder	1″x4″x10′	2
Rafters	2″x4″x10′	56	Feeder	1″x8″x10′	2
Sheathing	1″x6″	3500 Lin. ft.	Feeder	1″x8″x10′	2
Shingles	....	18,500	Feeder	1″x13″x14′	1
Studs	2″x4″x12′	35	Feeder ship-lap	....	94 Bd. ft.
Studs	2″x4″x14′	28	Finishing	1″x4″x14′	5
Studs	2″x4″x22′	4	Ridge Board	1″x4″x16′	43
Studs	2″x4″x10′	7	Corner Bds., etc.	1″x4″x12′	2
Studs	2″x4″x16′	7	Corner Bds., etc.	1″x4″x10′	2
Purlin	2″x4″x22′	20	Corner Bds., etc.	1″x6″x16′	8
Sills	2″x4″x16′	16	Corner Bds., etc.	1″x10″x12′	10
Rafter Br.	2″x6″x14′	28	Nests	2″x4″x12′	3
Sideing	1″x6″	1350 Bd. ft.	Nests	2″x2″x12′	5
Roost and)	2″x4″x18′	14	Nests	1″x3″x12′	29
Dropping)	2″x4″x20′	9	Nests	1″x4″x12′	17
Boards)	1½″x2½″x16′	15	1 window 3′x4′		
Roost and)	1½″x2½″x20′	3	3 doors 3′x6′-6″		
Dropping)	1″x4″x12′	3			
Boards) ship-lap	....	500 Bd. ft.			
Feeder	2″x4″x12′	5			
Feeder	2″x4″x10′	2			

BILL OF MATERIAL

Hardware

Nails—20d—40 lbs.
  8d common 75 lbs.
  8d Case—25 lbs.
  Shingle—116 lbs.

Bolts—½"x6" Anchor bolts—40
  ½"x4½" for feeder—10
  ½"x4½" eye bolts for feeder—4

Hinges—16 pair—3"—for ventilating boards
  10 pair 6" for roost
  17 pair 2½" butt hinges for nests
  1 pair 2½" butt hinges for feeders

Chain—75' light chain for roost
  8' (3/16" wire) chain for feeder

*Wire netting—1" mesh 4'x96' for open front
  Galv. wire fruit drier cloth—¼" mesh 1'x100'

Floor
  Concrete work (1:2.5:5 mixture. 1:2 wearing surface)
  Cement—35 bbl. Portland
  Sand—12.5 cu. yd.
  Stone or Gravel—25 cu. yd.

---

*Use graduated mesh wire with 4" mesh at bottom if moist mash trough is used.

Note: This bill does not contain material for windows in rear or any fixtures in feed room.

www.ingramcontent.com/pod-product-compliance
Lightning Source LLC
Chambersburg PA
CBHW082150230526
45467CB00043B/2790